MODERN
NUMEROLOGY

ANNE CHRISTIE

Published in 2001 by Caxton Editions
20 Bloomsbury Street
London WC1B 3JH
a member of the Caxton Publishing Group

© 2001 Caxton Publishing Group

Designed and produced for Caxton Editions
by Open Door Limited
Rutland, United Kingdom

Editing: Mary Morton
Coordination and Typesetting: Jane Booth
Digital Imagery © PhotoDisc Inc.

All rights reserved. No part of this publication may be reproduced or transmitted in any form or by any means, electronic or mechanical, including photocopying, recording or any information storage and retrieval system, without prior permission in writing from the copyright owner.

Title: Numerology
ISBN: 1 84067 280 3

MODERN

NUMEROLOGY

ANNE CHRISTIE

CAXTON EDITIONS

CONTENTS

6   INTRODUCTION

10   THE NUMERICAL ALPHABET

14   SINGLE NUMBER INTERPRETATIONS

32   COMPOUND NUMBER INTERPRETATIONS

CONTENTS

46 USING DIAGRAMS FOR INTERPRETATIONS

HOW TO USE NUMEROLOGY

WHAT IS A LIFE NUMBER?

WHAT ARE PERSONALITY NUMBERS?

HEART NUMBERS

DESTINY NUMBERS

KARMIC NUMBERS

76 DAYS, MONTHS AND YEARS

MONTHLY KEYWORDS

94 INDEX

Right: many early civilisations, including the Egyptians, studied numbers as a science and recorded their discoveries.

Below: numbers are universal and no-one knows their origin. It has been said that everything has been hidden in numbers since the beginning of human awareness of time.

INTRODUCTION

WHAT IS NUMEROLOGY?

Numerology is a science and art, which is little understood and often much maligned. To many people, numerology is simply a fun game, while to others it is a subject to be studied seriously. Numbers have held a special fascination for many thousands of years. It has been the means for generations of people to try to understand themselves and unravel the mysteries of the future. Numbers are universal and no-one knows their origin. We do know that many early civilisations such as the Egyptians, the Hebrews, the Chaldeans and the Hindus studied numbers as a science and recorded their discoveries. It has been said that everything has been hidden in numbers since the beginning of human awareness of time.

INTRODUCTION

The past and the future can be linked through numbers rather than language. Numbers are a universal way of communicating – they have a language of their own, and in numerology it is the meaning of each number and the energy it represents which matters.

It has been recorded that around 10,000 years ago the ancient Babylonians and Egyptians first used numbers scientifically. Traditional interpretations of each number have been traced back to the very dawn of civilisation.

Numerology has always had a mystic or spiritual significance.

The two main sources of current systems are the Greek philosopher, mathematician and astrologer, Pythagoras, born about 580 BC, and the Hebrew Kabbala, ancient Jewish tradition based on the Old Testament.

Left: numbers are a universal way of communication which links our past to our future.

INTRODUCTION

Right: when Pythagoras returned home to Greece he established a school of philosophy to which he admitted only a few students who were sworn to secrecy about what they learned.

PYTHAGORAS

Pythagoras is sometimes called the Father of Numerology. A large part of his working life was devoted to the study of numbers. He was convinced that numbers had mystical properties and he used a system of classification. In school today we still learn the geometrical theories formulated by Pythagoras.

Pythagoras may have first become fascinated with numbers during the time he lived in Egypt. When he returned home to Greece he established a school of philosophy to which he admitted only a few students who were sworn to secrecy about what they learned. We now know that Pythagoras believed that each number had its own unique personality, that some numbers were more powerful than others and that numbers contained the secrets of the entire universe. He suggested that the most powerful numbers were the odd, masculine numbers and the even, less powerful numbers were feminine.

TRIANGULAR NUMBERS

1 3 6 10

SQUARE NUMBERS

1 4 9 16

INTRODUCTION

He was a man far ahead of his time, believing that numbers had enormous cosmic significance, and he spent much of his life attempting to connect philosophical thought with mathematics.

It was Pythagoras who taught about the effects of music on the mind and the importance of meditation. The link which Pythagoras made between music and mathematics was expanded in the 17th century by the astronomer Johannes Kepler.

The classification of numbers which Pythagoras worked with separated numbers into those which could be represented by triangles, and those which formed squares.

The system used in this book is based on the Western alphabet because the 22 letters of the Hebrew alphabet do not correspond exactly and need to be transcribed for interpretation.

Each number has its own special energy and vibration. Systems of interpretation are many and there is no one true system. It is the hidden meaning of the numbers which the study of numerology reveals.

FAMOUS NUMEROLOGISTS

During the Renaissance Henry Agrippa, the philosopher, devised a system which related man to numbers; Count Cagliostro, who was also known as Giuseppe Balsamo, invented a system of numbers which gave prophetic readings and both these systems are based on the ancient Kabbalistic system of numerology. It was believed that when the

letters of a person's name were transcribed into numbers, the results were similar to an astrological chart.

In the 20th century, one of the most famous numerologists was Cheiro. He was said to have had many famous and influential clients including King Edward V11. Cheiro, however, frequently used numerology to predict death, and the general opinion today is that this is both unsafe and unwise – as in all systems of prediction.

Below: Pythagoras made a link between music and mathematics which was expanded in the 17th century by the astronomer Johannes Kepler.

THE NUMERICAL ALPHABET

Right: you can learn a great deal about yourself, family and friends. It is fun to use, very simple, yet scientific.

HOW DOES NUMEROLOGY WORK?
We use what is called a number alphabet.

In this book, the numerical alphabet is based on the Pythagorean or Western system. This is because the Kabbalistic system excludes the number 9. The Hebrews believed that this was the number which represented God and should not be used.

This is the tool kit you will need to use numerology to look at your own life, and begin to work out character profiles for others.

It needs no extra talent or psychic skill and you don't have to be a

mathematician to understand it. You will not have to memorise or use any complicated formulae or equations.

You can learn a great deal about yourself, family and friends. It is fun to use, very simple, yet scientific. For your own personal growth, you may be able to pinpoint areas for change in your life.

THE ALPHABET OF NUMBERS USED IN THIS BOOK

```
1 2 3 4 5 6 7 8 9
A B C D E F G H I
J K L M N O P Q R
S T U V W X Y Z
```

THE NUMERICAL ALPHABET

How can I start using numerology?

By using keywords which are related to numbers and specific energies.

1	PERSONAL RESOURCES
2	PERSONAL FEELINGS
3	PERSONAL CREATIVITY
4	INSTINCT AND LOGIC
5	EXPANSION AND SENSES
6	INTUITION AND THEORY
7	SETTING LIMITS
8	TRANSFORMATION
9	SPIRITUAL CREATIVITY

Does numerology have anything to do with colour?

It is generally accepted that energy is associated with colour.

1	THE PHYSICAL BODY
2	THE EMOTIONS
3	CREATIVITY
4	THE MATERIAL WORLD
5	INTELLECT
6	EFFECTIVENESS
7	COMMUNICATION
8	SPIRITUALITY

Below: it is accepted that energy is associated with colour and we see it daily as the sun rises and falls, but does numerology have anything to do with colour?

THE NUMERICAL ALPHABET

What about numerology and the planets?

The planets are linked to numerology in the following way:

Number		Planet	Sign	Keyword
1	Ego	Sun	Leo	Positive
2	Caring	Moon	Cancer	Feelings
3	Action	Mars	Aries	Initiative
4	Instincts	Mercury	Gemini/Virgo	Thought
5	Learning	Jupiter	Sagittarius	Expansion
6	Imagination	Venus	Taurus/Libra	Discrimination
7	Time	Saturn	Capricorn	Wisdom
8	Material transformation	Pluto	Scorpio	Subconscious
9	Karma	Neptune	Pisces	Spirituality
0		Uranus	Aquarius	Originality

Do numbers have different types of significance?

There are masculine and feminine numbers – first suggested by Pythagoras.

From 1–9 the numbers characterise active and extrovert characteristics – masculine energy – or passive and introvert characteristics – feminine energy.

These are not sexual associations, but relate to the dynamic forces governing the energy of the number.

Masculine numbers

1 3 5 7 9

Feminine numbers

2 4 6 8

THE NUMERICAL ALPHABET

IS THERE ONE CORRECT WAY TO BEGIN USING NUMEROLOGY?
There are various ways of using the basic knowledge:

LIFE NUMBER
is the personal number which is derived from the sum of the digits of the birth date. This is the most prominent number which stays with you throughout your life. It is normally reduced to a single digit.

NAME OR PERSONALITY NUMBERS
are now generally agreed to have less significance to the individual. It is impossible to change a birth date, but your name is chosen for you by your parents and can be – and often is – changed. They should not be dismissed, though, because the name number added to the life number completes any analysis of the personality.

HEART NUMBER
is also sometimes known as the soul number. This number will indicate your feelings and what you want from your life. It does not mean that other people will see this aspect of your personality.

DESTINY NUMBERS
show how we behave and often relate to the names which are given to us by others rather than the names we choose. These numbers are important in any numerological assessment and will frequently include the compound numbers.

KARMIC NUMBERS
are sometimes termed Fardic numbers. This emphasises the karmic path of a person – what should be learned in the present life.

Above: numerology can help us understand more about ourselves and help us to make decisions about our future. Our whole existence can be linked to numbers.

SINGLE NUMBER INTERPRETATIONS

How can numerology affect my life?
The best way to begin is to study the general interpretations of each number.

Interpreting single numbers
Single numbers represent the personal or life number.

Number 1:
Vibrates to the Sun and represents benevolence, creativity and protection. It is the number of original action and is the initiating basis of all other numbers. It is symbolic of the essential being of the individual.

Keywords:
Ego
Leadership
Personal identity

Sun Sign:
Leo

Symbol:
The Lion

Any person born on the 1st, 10th, 19th or 28th day of any month in any year is influenced by the number 1. Those born on 10th, 19th or 28th day of the month will in addition be challenged by these compound numbers. This is also true of the single and compound number of the name of the person.

This individual possesses a marked dislike of criticism and a strong sense of self-worth, demanding and usually getting respect from others. This person will insist on controlling and organising everyone and everything.

Their underlying desire is to be original, creative and inventive. These are people whose views are quite definite and, when thwarted, they can be stubborn. Whatever they undertake they will nearly always rise to a position of authority. Usually they insist on being looked up to by everyone – friends and family, colleagues and even the boss. If not, they will become very frustrated and will take it out on everyone in sight.

Number 1 will take on others' burdens, protect the weak and defend the helpless as long as the others do exactly as number 1 tells them. They always know better than anyone else, and are quite certain that their opinions are correct. Most of the time they are right which is annoying to everyone else – especially those who are expected to listen to their lecturing.

SINGLE NUMBER INTERPRETATIONS

Left: number 3 people will often find a career in the military or government departments.

They have a strong tendency to defend the underdog in any dispute and they show the same fierce loyalty to their pets and to all animals.

Marriage and family ties will only work successfully when the number 3 person has total freedom. This is associated with gambling, tests of physical strength, and risk-taking in all its forms; these people will bet on anything.

However, their optimism is contagious and their delightful bubbly personalities irresistible.

They are a strange blend of the happy-go-lucky clown and the wise philosopher – together with a marked lack of any sense of responsibility. Sometimes their ambitions, dreams and goals are serious and attainable – sometimes they are just frivolous and silly.

HEALTH:
Sciatica or skin complaints; nervous problems through overwork.

WORK:
The freedom to travel, a sense of authority, government or military work.

SINGLE NUMBER INTERPRETATIONS

NUMBER 4:
Represents practicality, the instincts, logical thought, the concrete and the material world. Number 4 knows how to use intellectual energy effectively and vibrates to the planet Mercury.

KEYWORDS:
Instinct
Logic

SUN SIGNS:
Gemini/Virgo

SYMBOLS:
The Twins/The Virgin

A person born on the 4th, 13th, 22nd or 31st day of any month in any year is a number 4. Those born on the 13th, 22nd and 31st have the challenge of their compound number throughout their lives. This is also true of the name of the individual.

All number 4 people are misunderstood by their families and friends. They make their own enigmatic rules, which don't always correspond to those of society.

Every thought and action is marked with a peculiar individuality. They frequently shock others by their actions and speech – often it would seem to be deliberate. If there is a different way to do anything, the number 4 person will do it.

These are people who care little for the present; their concern is for the future and they are indeed frequently light years ahead of the rest of us. They seem to have an inborn talent for prophecy – knowing instinctively what will happen or be fashionable long before it arrives.

Theirs is usually an unconventional lifestyle, although more often than not their crazy ideas turn out to be surprisingly successful. The curious nature of these individuals draws them to anything off the beaten track and any incredible unscientific or unproven theory excites the number 4 person to convince himself and everyone else that the most way-out notion can be turned into reality.

SINGLE NUMBER INTERPRETATIONS

To tell a number 4 person that something is impossible simply spurs him on and intensifies his urge to prove them wrong. There is no such phrase as "Mission Impossible" to a number 4.

The vibration of number 4 is for change in every area of life, but these people are usually reluctant to ever change their own personal habits, They can be both fixed and stubborn when urged by others to become more socially acceptable.

Friendships are vital – money has little or no meaning. There is not the slightest desire to impress others and they don't care where they live as long as they can keep their imaginations intact.

They genuinely fail to notice their surroundings and they believe in live and let live, however outrageous this may turn out to be. In return they expect to be given the same consideration by others.

HEALTH:
Kidney or bladder disturbance, headaches, nervous tension.

WORK:
Building, architecture, design in traditional organisations.

Below: a number 4 person has the ability to convince himself and everyone else that the most way-out notion can be turned into reality. This is reflected in the occupation that number 4 people choose, often in building, architecture and design.

SINGLE NUMBER INTERPRETATIONS

NUMBER 5:
Represents expansion, flexibility, tolerance, the senses, communication, versatility and movement. It rules learning and vibrates to the planet Jupiter.

KEYWORDS:
Expansion
Tolerance

SUN SIGN:
Sagittarius

SYMBOL:
The Archer

The number 5 influences a person if they are born on the 5th, 14th or 25th day of any month of any year. The compound numbers of 14 and 23 challenge people born on those days. This is also true of the number of their names.

Number 5 people are innately courteous and charming, but if they spot flaws and mistakes they will not hesitate to point them out. These people are supercritical and cannot ignore their own or others' mistakes.

They love travel and movement – change is a necessity to them in all areas of their lives.

Sometimes they will over-analyse situations and people and it is hard for them to rely on their intuition and feelings. This can be the cause of breakdowns in all their relationships – no partnership can bear the pressure of constant scrutiny. These are people who can talk a love affair to death. Logic has nothing to do with love.

They are good to spend time with, as outwardly they are pleasant and amenable. Number 5 vibrates to the higher intellect and these are extremely bright people – mentally alert with higher than average intelligence. They miss nothing and are fine-tuned to pick up on the tiniest detail. They know instinctively how to use the space they inhabit.

If money pressures prevent these individuals from travelling frequently, they will daydream and travel in their minds equally well – they have vivid enough imagination to satisfy their wandering urges.

Traditionally earth magic is associated with the number 5 – and these people often have a fascination with fairies, magic and anything mysterious. Their need to analyse every last detail is at odds with this and these people frequently find it hard to understand themselves.

SINGLE NUMBER INTERPRETATIONS

5

Number 5 people are usually highly strung, craving excitement and living on their nerves. Often they act impulsively with the ability to think and act quickly. They are speculators at heart, with the keenest sense for new inventions and the willingness to take risks. Successful occupations are writing, publishing, advertising and public relations. They have a flexibility of view and the ability to recover from bad experiences – which seem to leave little or no long-lasting impression on them.

HEALTH:
Insomnia, neuritis, nervous problems.

WORK:
Sales work, teaching, publishing, speculation, travel.

Left: number 5 vibrates to the higher intellect and these are extremely bright people – mentally alert with higher than average intelligence. They miss nothing and are fine-tuned to pick up on the tiniest detail; often finding work in the teaching profession publishing or speculation as a result.

NUMBER 6:

Represents feminine qualities. It is associated with abstract thinking, fantasy, intellectual creativity and the imagination. It vibrates to the planet Venus.

KEYWORDS:
Imagination
Theory
The abstract

SUN SIGNS:
Taurus/Libra

SYMBOLS:
The Bull/The Scales

An individual is influenced by number 6 if they are born on the 6th, 15th or 24th day of any month in any year.

These people seem to have a magnetic attraction for others. Their families, friends and associates genuinely love them. When they form attachments to others, they are devoted to those they become close to. Their nature is motivated by affection and idealism rather than sex and they are born romantics, with a streak of sentimentality which is impossible for them to deny or hide.

They love music, art, beautiful homes with the most tasteful furnishings, and harmony in all their surroundings.

Entertaining friends and making others happy come high on their list and they cannot face jealousy, discord, arguments or unpleasantness of any sort. Number 6 people make friends easily, enjoy settling the disputes of their relatives and friends and appear to be the most docile and amenable people – until crossed. When their stubborn nature kicks in, they are not quite so appealing.

Sometimes, through their own abilities and talents, and sometimes through inheritance, money often comes without effort. But they can veer between extravagance and stinginess.

SINGLE NUMBER INTERPRETATIONS

Most number 6s love the country – they find it emotionally soothing to live or spend time near trees or water. They have a marked fondness for luxurious things and find anything ugly very offensive. They hate vulgarity or loud behaviour and have the highest admiration for anything tasteful and refined.

As a rule these people have impeccable manners and are always polite. However, when they have strong feelings about anything, they do not hesitate to make their opinions quite clear. Debate and discussion come easily to them and, with their logical minds, they usually win others over to their point of view. These people have the most irresistible smile.

HEALTH:
Throat, voice, nose, lungs, circulatory problems. Lack of exercise causes weight gain.

WORK:
The arts, entertainment industry, health – especially diet.

Above: *most number 6s love the country – they find it emotionally soothing to live or spend time near trees or water. Their love of art can also be reflected in the occupation they choose.*

SINGLE NUMBER INTERPRETATIONS

NUMBER 7:
Represents time, setting limits, the boundaries of the material world, material attachments, the bridge to the spiritual world. It vibrates to the planet Saturn.

KEYWORDS:
Time
Boundaries
Setting limits

SUN SIGN:
Capricorn

SYMBOL:
The Water Goat

An individual born on the 7th, 16th or 25th of any month is influenced by this number – birthdays on 16th or 25th have the additional challenges of their compound number.

Number 7 people often have amazing dreams, and sometimes they tell others about them. Sometimes they just keep quiet about them. They have a secret interest in the unknown – mythology, mysteries or UFOs. Often these people are intuitive and clairvoyant with magnetic, calming personalities.

These are individuals who often have strange and unorthodox ideas – about religion, politics or just life in general; they object to following the herd and sometimes believe in the newest religious cults. This is the number which indicates the totality of the material world.

Number 7 people will probably become seasoned travellers, reading travel books and soaking up information about foreign countries and people. Many of them will find they have a strong attraction to the sea, either water sports, sailing or even joining the navy.

They can become very anxious about the future and need to know that they have some financial security to protect them. Actually, they don't have much time for material possessions, or even the accumulation of wealth, because they often find they can earn substantial sums of money. Their original ideas and creativity encourage them towards the arts – writing, acting, singing or dancing. They may make significant contributions to charities in a quiet, very laid-back way.

SINGLE NUMBER INTERPRETATIONS

Number 7s are unlikely to talk much about their ambitions and everything they do may have a slightly philosophical tinge. Friends and family find themselves with a champion who is always a sympathetic and understanding listener. However, these people always keep their own problems private and do not unburden themselves to others. They never ask prying questions or invade others' privacy – they have a horror of this happening to themselves.

A sensitive nature, artistic temperament and the most refined manners front a sometimes surprisingly taciturn mind. They are truly unprejudiced and non-judgmental. It is hard to get a number 7 person to tell you what he or she is really thinking – they live in a Neptunian world of secret dreams which they will keep secret unless and until they really trust you.

Seven is the symbol of Kronos, the Greek god, which reminds us that everything of limited duration is not real. Number 7 symbolises totality in the material world – seven notes in the musical scale, seven colours of the rainbow, seven virtues, seven deadly sins, seven chakras to the body, seven planets known to the ancient world, seven days of the week. Seven is the number which tempts people to try and control time.

HEALTH:
Skin complaints, bad diet and worry will cause problems.

WORK:
Looking after others; artistic or literary career, TV, films, travel, especially connected with the sea.

Below: many number 7 people find they have a strong attraction to the sea and take up water sports, sailing or even join the navy to fulfil this passion.

SINGLE NUMBER INTERPRETATIONS

NUMBER 8:
Represents doing what has to be done, the unconscious mind, transformation of the material, balance, timeless space. It vibrates to the planet Pluto.

KEYWORDS:
Wisdom
Material transformation
Unconscious mind

SUN SIGN:
Scorpio

SYMBOL:
The Scorpion/Eagle/Phoenix

Someone born on 8th, 17th or 26th day of any month is ruled by the number 8.

Number 8 people are reserved and quiet. Although they are not pushy, they get where they want to by slowly and surely achieving their ambitions. If they appear to be shy, this is a cover for their intense drive to get to the top in their career.

These people understand about material reality – which is not real for very long. Money, sex and power are the areas in which number 8 operates. The necessity for trust gives number 8 its strange, mystic reputation.

Number 8 individuals are quite happy to wait, using their time wisely, for their plans to bear fruit and it's unusual to find them procrastinating. Their sense of responsibility and duty forbids time-wasting. Normally they will do just what is expected of them. A good sense of humour, which is subtle and dry, makes them easy people to be with. They behave as though they couldn't care less what anyone else thinks of them, but in reality they secretly enjoy compliments.

Most number 8 people have enormous inner strength and very deep intense natures. Often they have an important role to play in the lives of others and sometimes they can be fanatical about religion. Although they can make many loving friends, they can also acquire bitter enemies.

SINGLE NUMBER INTERPRETATIONS

They often seem to be undemonstrative and cold even with those they love and trust. They show affection shyly and are devoted to their friends and family. Loneliness, and a desperate need to be loved, lies underneath the surface. If necessary, they can go to great lengths to make sacrifices for those who depend on them or for a strongly held ambition or ideal.

As they grow older and mature, they seem to look and behave younger than their years. They demand much of themselves and others and, for all their appearance of discipline, maturity and self-control, at heart they are needy and lonely. The wholehearted pursuit of happiness is hard for them.

HEALTH:
Liver problems, rheumatism, headaches, diseases of the blood.

WORK:
Public life, large organisations, fund-raising for charity, any kind of business.

Left: number 8 people often find careers in public life or large organisations.

Below: number 8s demand much of themselves and others and, for all their appearance of discipline, maturity and self-control, at heart they are needy and lonely.

SINGLE NUMBER INTERPRETATIONS

NUMBER 9:
Represents divine love, completion, innate talents, spiritual creativity and karmic reward, and vibrates to the planet Neptune.

KEYWORDS:
Completion
Spirituality
Karma

SUN SIGN:
Pisces

SYMBOL:
The Fishes

Those who are born on the 9th, 18th or 27th day of any month are number 9 people.

This is the number of anti-ego and divine love. These are people who are determined to get their own way, although they are not stubborn by nature. They tend to make snap decisions and can be impulsive – often regretting their actions later. Number 9 people have a tendency to flare up at others frequently although they will quickly forgive and forget.

They are trusting by instinct, which can lead them into trouble. Because they are direct and straightforward, they expect others to be the same and are often disappointed. When people are manipulative or devious they are always shocked – generally they are quite incapable of such behaviour themselves, and they can be caught off guard by dishonesty. They need to learn to be more cautious and less trusting. Number 9 indicates understanding, compassion and the highest ideals of selfless love.

These people have an amazing ability to go straight to the heart of a situation, bypassing the need for lengthy analysis. Slow thinkers drive them wild – and their impatience with others does not win them many friends or allies at times.

There is a complete lack of guile, and what you see in a number 9 person is exactly what you get. They never play games to get what they want – they simply demand it. Others find this so surprising that they often capitulate.

A childlike quality of vulnerability is touching for others to see, and number 9 people often find others protecting them. Others find this foolish and find it hard to respect them – until they experience the courageous spirit and violent temper. Then they think again.

SINGLE NUMBER INTERPRETATIONS

Number 9 people often appear to be vain – they are not really, although they are very concerned about their own appearance. A lack of confidence and fear of rejection is the cause of this. Although they may appear to be assertive, independent and pushy, they need the constant reassurance that they are loved, liked, admired and respected.

They are often very extravagant and very generous to others; instinctively they will let tomorrow take care of itself, let go of everything and just give. The number 9 tunes in to happiness and joy.

HEALTH:
Fevers. Avoid rich and spicy foods.

WORK:
Philanthropic work, art, writing, music, religious work, theatre, entertainment, military. Excellent advisers.

Above: the number 9 tunes in to happiness and joy; music and entertainment often being closely related to the work they choose.

COMPOUND NUMBER INTERPRETATIONS

Right: number 11 represents conflict which needs to be resolved in some way.

Below: the wheel of fortune illustrates well the number 10 as it is is the number of rise and fall according to the chosen action – for good or evil.

Compound numbers represent personal karma.

10 The Wheel of Fortune

The number of rise and fall according to the chosen action – for good or evil.

Arouses extreme responses – respect, hate or fear.

Everything is self-determined.

The power to turn creative concepts into reality – must be used wisely to avoid absolute destruction.

Compassion and self-discipline must be maintained at all times.

Avoid feelings of frustration, and proud or arrogant behaviour.

11 Clenched Fist

Hidden trials and treachery from others.

Two opposing situations or individuals are represented.

Difficulties arise from the illusion of separation.

Unite divided aims and avoid a third interfering force.

Any separating force must be identified and attempts made to achieve compromise.

Conflicting desires within the self create a mirror and reflect the problem.

Two desires or forces, which are distinct, may unite for happiness whilst always retaining their individuality.

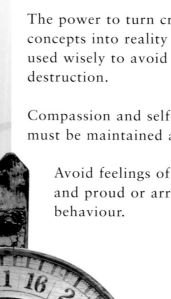

COMPOUND NUMBER INTERPRETATIONS

Left: symbolically the number 13 is represented by a skeleton or death.

12 THE VICTIM

Periodic sacrifice for the plans of others.

Be alert to every situation; beware of false flattery.

Take care that others do not seek to use you for their own gain. Suspect those who offer high office or position.

Forewarned is always forearmed.

Personal goals may need to be sacrificed to others ambitions.

This number represents the process of education – the student and the teacher.

Sacrifices may be needed to achieve intellectual and spiritual wisdom.

Always look within for solutions.

Pay attention to education – this will bring success and end suffering.

13 CHANGE

Not an unlucky number.

He who understands 13 will be given power and dominion.

Symbolically a skeleton – or death – with a scythe, reaping down men in a field of new-mown grass. Young heads and faces thrust through the ground and emerge all around.

The number of upheaval to break new ground.

Associated with power – used selfishly will bring personal destruction.

A warning of the unexpected and unknown.

Adapt to change gracefully – use the strength of this vibration and decrease the negativity.

Associated with explorers, new discoveries and genius.

COMPOUND NUMBER INTERPRETATIONS

14 Challenge

Communication with the public associated with writing, publishing, the media.

Beneficial changes in partnerships of all kinds and business.

Speculation brings luck.

Movement and travel involving people and nations can be fortunate.

Gains and losses may be temporary.

Currents of change are always present.

Danger of accidents from natural causes – elements of fire, earthquakes, floods, tornadoes.

It is risky to depend on the word of those who may be in a position to misrepresent situations – don't rely on others.

Rely on the self, intuition and the inner voice.

Money dealings and speculation can be lucky – danger of loss is always present through taking wrong advice or over-confidence

15 The Magician

Number of deep significance – an alchemical vibration which manifests magic.

A lucky number which can bring blessings – the ability to bring happiness to others and shine light into darkness.

Associated with good talkers and musical or dramatic ability.

Personal magnetism and a dramatic temperament; great charisma.

Fortunate for obtaining gifts, favours from others and money.

Rules the lower levels of the occult when associated with the numbers 4 or 8.

Can be associated with black magic as a victim.

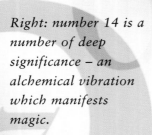

Right: number 14 is a number of deep significance – an alchemical vibration which manifests magic.

COMPOUND NUMBER INTERPRETATIONS

16 Shattered Citadel

The tower struck by lightning was the ancient image associated with this number.

Warning of strange fatality, the defeat of plans and accidents.

Make all plans in advance.

Anticipate the possibility of failure and circumvent it by paying careful attention to detail.

Listen to the voice within which will warn of danger in dreams in time to avoid it.

Finding success and happiness in ways other than leadership, fame or celebrity will decrease the negative aspects of 16.

17 The Star

The symbol of the eight-sided star of Venus associated with the ancient magi. An image of love and peace which promises a spiritual rise above trials and tribulations with the ability to overcome former failures – both in personal relationships and career.

The number of "immortality".

A fortunate compound number.

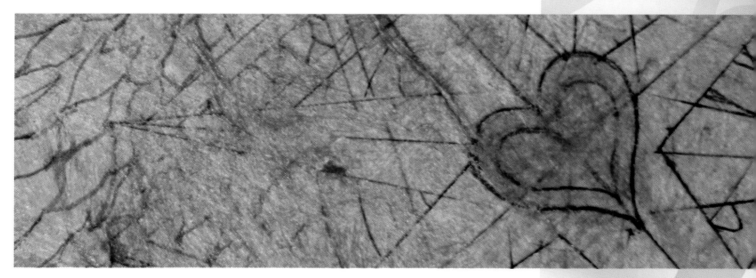

Below: number 17 is represented by the eight-sided star of Venus associated with the ancient magi – an image of love and peace.

COMPOUND NUMBER INTERPRETATIONS

Right: number 19 symbolised by the sun.

Below: number 18 warns of danger from the elements – electric shocks or lightning, fires, floods earthquakes, explosions.

18 Conflict

This is symbolised as materialism striving to destroy spirituality.

Sometimes represents bitter quarrels – social revolution and upheaval, wars and family disputes.

The acquisition of money or position through conflict.

Warns of treachery and deception from friends and enemies. Danger from the elements – electric shocks or lightning, fires, floods earthquakes, explosions.

Always meet deception or hatred from others with generosity and kindness, forgiveness and love.

This is the level of a choice made by the higher self between incarnations to test the soul.

Love always conquers conflict.

19 The Sun

Symbolised as the Prince of Heaven, this is one of the most fortunate of the compound numbers.

It indicates victory over all earthly failure and disappointment.

Blesses and promises fulfilment and happiness with personal success. All ventures will smooth the path and dilute any negative vibrations.

COMPOUND NUMBER INTERPRETATIONS

20 JUDGEMENT
Pictured as a winged angel sounding a trumpet to a man, woman and child arising from a tomb with their hands clasped in prayer.

There will be a powerful awakening some time bringing new ambition, new purpose or new plans.

A call to action.

Delays will be conquered by cultivating faith in the personal power to transform.

Precognitive dreams.

Not a material number – financial success is not seen.

Little care for financial matters if the basic necessities are provided for.

21 THE UNIVERSE
The crown of the magi promises success, advancement, awards and honour in the career and life in general.

Victory after a long struggle.

Soul testing and tests of determination.

Certainty of final victory over all opposition.

A fortunate number.

A number of karmic reward.

Above: number 21 is a fortunate number promising success and victory over all opposition.

COMPOUND NUMBER INTERPRETATIONS

22 Caution

Symbolised as a good man who has been blinded by the folly of others, carrying a knapsack on his back, full of errors.

He appears to be unable to defend himself against a fierce tiger which is about to attack him.

A warning of delusion and illusion. Indicates a good person who lives in a fool's paradise.

A dreamer of dreams only waking when surrounded by danger.

Warning of mistakes in judgement and putting faith in others who are untrustworthy.

Exercise caution; be watchful in personal and career matters.

Curb spiritual laziness and be more alert.

Develop more spiritual aggression.

Realise that you have the power to change things and prevent failure by focusing on success.

Be in control of events.

23, 32 Royal Star of the Lion

A promise of success in career and personal affairs.

Karmic reward numbers.

Guarantee protection from those in high places and help from superiors.

Most fortunate numbers.

No other numbers can challenge the strength of the lion and expect to win.

Right: 23 and 32 promise success in career and personal rewards; most fortunate numbers.

COMPOUND NUMBER INTERPRETATIONS

24, 33 CREATIVITY

Numbers of karmic reward.

Promise the assistance of those who have power.

Indicate close associations with those of high position and rank.

Increase financial success.

Great ability to achieve happiness in love.

Magnetism is very attractive to the opposite sex.

Gain through love, the arts or law.

Be warned about self-indulgence and arrogance in career, finances and love.

Be sure to appreciate the benefits of these compound birth numbers – do not let good fortune make you careless about spiritual values.

Do not become selfish.

There may be a temptation to overindulge.

25, 34 ANALYSIS

Spiritual wisdom is gained through the careful observation of people and things.

Worldly success is achieved by learning from experience.

Overcoming disappointments early in life brings strength.

Learning from past mistakes is a rare achievement.

Excellent powers of judgement.

Not material numbers – financial benefits of any great substance will not be forthcoming.

Above: 25 and 34 are all about spiritual wisdom; overcoming disappointments in life brings strength.

COMPOUND NUMBER INTERPRETATIONS

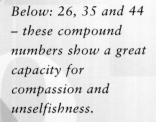

Below: 26, 35 and 44 – these compound numbers show a great capacity for compassion and unselfishness.

26, 35, 44 Partnerships

The ability to help others but not yourself.

Great capacity for compassion and unselfishness.

Many contradictions.

Warnings of failure, danger and disappointment brought about through bad advice.

27, 36, 45 The Sceptre

Fortunate number of courage and harmony touched with enchantment.

The promise of a person blessed with authority.

Productive work will bring great rewards from using the imagination and intellect.

Always carry out your own original plans and ideas.

Do not be intimidated or influenced by other opinions or by opposition.

A number of karmic reward.

COMPOUND NUMBER INTERPRETATIONS

28 THE LAMB

A number full of contradictions.

There is the promise of genius and the possibility of achieving success – which is often realised – and then it is all taken away, unless very careful provision has been made for the future.

There is loss through misplaced trust in others.

Powerful opposition from competitors in business and enemies.

Danger of serious losses through law suits.

The possibility of having to begin life again.

The key to this number is to look before you leap.

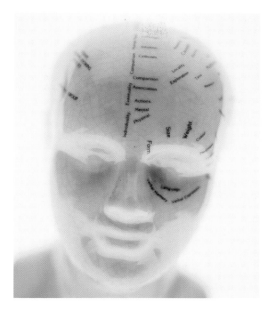

29, 47 GRACE UNDER PRESSURE

This number suggests a test for spiritual strength through a series of trials and tribulations.

Life is full of unreliable friends, treachery and deception, uncertainty, unexpected danger.

There is much anxiety and grief caused by those of the opposite sex. Grave warnings for each area of the career and personal life.

Development of absolute faith in the power and goodness of the self. Constant energetic cultivation of an optimistic frame of mind.

An acceptance of full responsibility and the ability not to blame others.

Left: 29 and 47 test all the emotions when they come up as compound numbers.

Below left: the lamb, symbolic of number 28.

COMPOUND NUMBER INTERPRETATIONS

Right: loneliness, isolation and self-containment are suggested by the numbers 31, 40 and 49.

Below: 30, 39 and 48 are the numbers of meditation and mental superiority.

30, 39, 48 MEDITATION
Mental superiority.

Introspection, thoughtful deduction and a desire to put all material things aside

Neither fortunate or unfortunate, this number depends on the wish of the person it represents. Few can be relied on as friends. This is the number of a loner who does not enjoy public gatherings or social functions.

Mental energy directed to developing something worthwhile – ideas which may change the world or protecting and developing personal talents like art or music.

Although this may be a lonely life, it can be infinitely rewarding.

31, 40, 49 THE HERMIT
These are similar to the previous numbers – and suggest even more loneliness, isolation and self-containment.

Often the self-sufficiency of high intelligence is present.

The world may be rejected for the sake of peace and quiet – or a retreat from society and its pressures.

Sometimes this is an opinionated person who advocates political change while refusing to alter anything on a personal level.

A sense of isolation and loneliness, even in a crowd.

COMPOUND NUMBER INTERPRETATIONS

Left: an emphasis on romance and love is associated with the compound number 33.

32, 41 COMMUNICATION

The ability to sway large crowds of people – similar to the powers of 14 and 23.

A natural ability to charm others sometimes indicates a political ambition.

Television, writing, publishing and media activities.

The ability to work under pressure A fortunate number for the person who holds onto their own judgement and opinions in all areas of life.

Plans could be wrecked by others' stupidity and stubbornness.

33 SENSITIVITY

Associated with an extremely sensitive nature.

Many fortunate friendships.

Strong magnetic appeal for the public.

Productive partnerships.

Emphasis on romance and love.

Unconventional sexual attitudes.

COMPOUND NUMBER INTERPRETATIONS

34, 25
Claimed to symbolise revolution, strife, conflict, war and upheaval.

Repeated failure and disappointment.

36 SEE 27

37 RELATIONSHIPS
This number is associated with a very sensitive nature; with good and fortunate partnerships, a very magnetic personality – particularly in a creative or media environment, and very productive partnerships, both business and personal.

It places great emphasis on romance and love – sometimes there is too much on sexuality. There is a tremendous need for harmonious relationships.

It is easier to achieve success and happiness when the individual is in a partnership than alone as an individual

Attitudes to sex may be unconventional

Right: 34 and 25 symbolise revolution strife, conflict, war and upheaval.

38	SEE 29
39	SEE 30
40	SEE 31
41	SEE 32
42	SEE 24
43	SEE 34
44	SEE 26
45	SEE 27
46	SEE 37
47	SEE 29
48	SEE 30
49	SEE 31
50	SEE 32
51	SEE 24

52 THE NATURE OF THE WARRIOR
A promise of sudden unexpected advancement in any undertaking.

Favourable for those needing military protection and any leader of a cause unrelated to war or conflict.

The threat of attempted assassinations and dangerous enemies.

USING DIAGRAMS FOR INTERPRETATIONS

Diagonal Line

1 5 9 — Communication

1, 5 and 9 indicates the importance of effective communication in all relationships – focusing on the awareness of self-identity and personal resources, the willingness to learn, expansion and the awareness of innate talents, spiritual creativity and the common source of all life.

3 5 7 — Effectiveness

3, 5 and 7 The energy in this line is that of effectiveness and administrative talents. The focus is on the ability to act, personal creativity, expansion, the senses, a willingness to learn and the consciousness of time – the need to set limits.

How does a number diagram work?

This shows how to work out basic numerology using diagrams and energy lines as a basis for interpretation.

left: 3, 5 and 7 reflect a willingness to learn and the consciousness of time – the need to set limits.

USING DIAGRAMS FOR INTERPRETATIONS

WHAT IS A LIFE NUMBER?

These are sometimes called birth path numbers. Many people will have the same life number and you may find that not everyone falls into the exact description given – these are general interpretations giving negative and positive aspects and you will find that people differ widely as you do more numerological analysis.

LIFE NUMBERS – HOW TO CALCULATE

Example birth date
Reduce to single and compound numbers
Reduce to compound and then single number
Keep reducing to single digit
If master numbers 11 and 22 appear, do not reduce them

HOW TO DISCOVER YOUR LIFE NUMBER:

1 Birth date: 7. 5. 1960   7+5+1+9+6
                              = 28
                              2+8
                              = 10

Compound numbers 28 and 10
Find interpretation of compound numbers 28 and 10.

2 Reduce to single number: single number = 1

3 Find interpretation of life number 1.

4 Circle number 1 on energy grid and find the keyword.

   This is your **life number** and the circled square on the grid shows your life energies.

   You can now synthesise the interpretations to make a complete analysis of the life number.

PAUL DICKERSON
Birthdate 27th October 1965.
27. 10. 1965

$2+7=9$
$1+0=1$
$1+9+6+5=21 \quad 2+1=3$
$9+1+3=13$
$1+3=4$

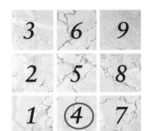

USING DIAGRAMS FOR INTERPRETATIONS

LIFE NUMBER 1

PERSONALITY KEYWORDS:
Leader, purposeful, tenacious (masculine)

ZODIAC ASSOCIATIONS:
Sun; Leo, Sagittarius

KABBALA:
Unity, wholeness totality

HEALTH:
Heart, circulatory problems, eye trouble, poor vision

WORK:
Creative with little restriction and opportunity of exercising authority.

LIFE NUMBER 2

PERSONALITY KEYWORDS:
Well balanced, sensitive, cheerful (feminine)

ZODIAC ASSOCIATIONS:
Moon; Cancer, Gemini, Libra

KABBALA:
Division, unity, relatedness

HEALTH:
Digestive system, stomach problems

WORK:
Teamwork, dealing with people, politics.

Left: life number 1 indicates a leader, possibly in a creative and non restricting work environment.

USING DIAGRAMS FOR INTERPRETATIONS

LIFE NUMBER 3

PERSONALITY KEYWORDS:
Versatile, adaptable, idealistic, charming

ZODIAC ASSOCIATIONS:
Mars; Aries

KABBALA: *Fertility, completion*

HEALTH:
Sciatica, skin complaints

WORK:
Freedom to travel but with potential for advancement.

LIFE NUMBER 4

PERSONALITY KEYWORDS:
Hard working, sensible, logical, materialistic

ZODIAC ASSOCIATIONS:
Uranus; Taurus, Virgo, Capricorn, Aquarius

KABBALA:
Solidity, reliability, the law

HEALTH:
Nervous tension, headaches, kidney/bladder problems

WORK:
Constructive, methodical, although within traditional organisation, unconventional reforming views.

Below: life number 3 people are versatile and adaptable who prefer a career where they have freedom to travel.

USING DIAGRAMS FOR INTERPRETATIONS

Left: life number 5 people are lively, creative and artistic; work in publishing, communications and teaching are often the choice for these people.

LIFE NUMBER 5
PERSONALITY KEYWORDS:
  Lively, creative, artistic
ZODIAC ASSOCIATIONS:
  Jupiter; Aquarius, Virgo, Gemini
KABBALA:
  Life, regeneration, creativity, expansion
HEALTH:
  Nervous problems, neuritis, insomnia
WORK:
  Communications, teaching, publishing.

LIFE NUMBER 6
PERSONALITY KEYWORDS:
  Balance, harmony
ZODIAC ASSOCIATIONS:
  Venus and the home
KABBALA:
  Fruitfulness, harmony, the home
HEALTH:
  Throat, nose, voice, lungs
WORK:
  Entertainment, health, the arts

USING DIAGRAMS FOR INTERPRETATIONS

Above: the ambitious and self sufficient number 8 prefers work in a large organisation, preferably in the financial world.

LIFE NUMBER 7

PERSONALITY KEYWORDS:
  Spiritual, sensitive, intuitive
ZODIAC ASSOCIATIONS:
  A magical number; links with Saturn, Neptune and the Moon
KABBALA:
  Mysticism, magic, spirituality
HEALTH:
  Skin and diet-related complaints
WORK:
  Caring, travel, literary/artistic work, TV/films

LIFE NUMBER 8

PERSONALITY KEYWORDS:
  Ambitious; self-sufficient, persistent
ZODIAC ASSOCIATIONS:
  Pluto, Saturn; both material and spiritual signs
KABBALA:
  Material concerns, leadership, justice, ambition
HEALTH:
  Blood disorders, liver complaints, rheumatism, headaches
WORK:
  In large organisations, excitement, challenge, financial world.

USING DIAGRAMS FOR INTERPRETATIONS

LIFE NUMBER 9
PERSONALITY KEYWORDS:
  Creative, spiritual, compassionate
ZODIAC ASSOCIATIONS:
  9 is the number of the universe; Mars; Aries, Pisces
KABBALA:
  Humanitarian issues, inspirational leadership
HEALTH:
  Fevers, rich food should be avoided
WORK:
  Theatre, the arts, religious organisation.

LIFE NUMBER 11
PERSONALITY KEYWORDS:
  Determined, highly principled, selfish, risk taker
ZODIAC ASSOCIATIONS:
  Master number; Neptune; Gemini This number should never be reduced.
KABBALA:
  No direct meanings
HEALTH:
  Guard against all over-indulgence
Work:
  Leadership, teaching.

LIFE NUMBER 22
PERSONALITY KEYWORDS:
  Balance, happiness; contentment
ZODIAC ASSOCIATIONS:
  The traditional number of perfection; the Sun
KABBALA:
  No direct meanings
HEALTH:
  Anxiety, depression, frustration
WORK:
  Vocational, non-materialistic, charitable.

Below: life number 11 people are highly principled and determined. Their work can include teaching and any occupation that requires leadership.

USING DIAGRAMS FOR INTERPRETATIONS

All the compound numbers show hidden aspects of the personality; they are sometimes referred to as spiritual or secondary numbers. In any analysis of the life numbers, compound numbers must always be considered even though the significance may be of secondary importance.

Compound numbers traditionally end at 52, suggesting that their power is associated with the weeks of the year. Many of the compound numbers have no zodiac association.

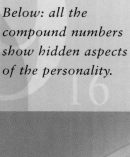

Below: all the compound numbers show hidden aspects of the personality.

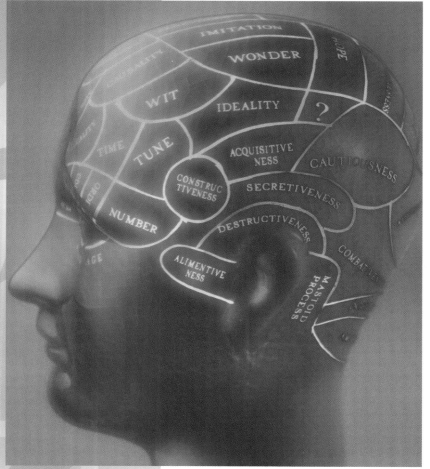

Compound Life Number 10
Linked to number 1. Linked to reaffirmation of this number. Attainment, faith, completion, fulfilment. Change is part of the life path.

Compound Life Number 12
The number of emotion and sacrifice, associated with the unseen and secrecy. Important to grasp any educational opportunities.

Compound Life Number 13
Linked to power and this is positive when used wisely. Change and rebirth, upheaval and trauma. The unknown. Important to remain adaptable.

Compound Life Number 14
Challenge and movement, fortunate for money, observe caution when taking risks.

Compound Life Number 15
Associated with magic. An element of power – luck and good fortune. A dramatic number.

USING DIAGRAMS FOR INTERPRETATIONS

COMPOUND LIFE NUMBER 16
Not usually considered a fortunate number. Often events will not go according to plan. A number which is passionate – volatile; people may be explorers, inventors or impulsive and rash.

COMPOUND LIFE NUMBER 17
Connections with intuition, a spiritual number. Sometimes called the number of immortality. Life may not be easy.

COMPOUND LIFE NUMBER 18
A struggle to balance the materialistic and spiritual sides of life. Often linked with the Moon and Cancer. Sometimes a difficult life path.

COMPOUND LIFE NUMBER 19
Success, good humour and happiness are associated with this number which is linked with the Sun and Leo. Good for speculation, recognition and dealing with children.

COMPOUND LIFE NUMBER 20
Linked to the moon, spiritual effort and fate. Life is never going to be boring or dull. The number of action, planning and new projects.

COMPOUND LIFE NUMBER 21
Ruled by the Sun; associated with power, success and achievement. Determination, material success and perseverance are connected with the benefits of karmic rewards.

COMPOUND LIFE NUMBER 23
With the help of others much will be achieved. A number of karmic reward with a special benefit of protection from anything harmful.

COMPOUND LIFE NUMBER 24
Linked to creativity, love and money. A fortunate number – happy relationships, charisma and personal magnetism. Aims will usually be achieved.

Left: number 23 compound life number is a number of karmic reward, with a special benefit of protection from anything harmful.

USING DIAGRAMS FOR INTERPRETATIONS

COMPOUND LIFE NUMBER 26
Partnerships of all description are the focus of this number; personal and professional alliances both successful and stressful. Tread warily and watch all relationships carefully.

COMPOUND LIFE NUMBER 27
Power and strength connected with intellectual matters. Originality and imagination will bring karmic rewards through hard work.

COMPOUND LIFE NUMBER 28
Associated with loss and the law. Trust is the keyword. Losses may arise through misplaced trust in business and personal affairs. A permanent battle for security.

COMPOUND LIFE NUMBER 29
Very much linked with karmic debts this number is associated with deception and treachery. Many hardships may be suffered. Friendships likely to be difficult.

COMPOUND LIFE NUMBER 30
An idealistic introspective loner is indicated by this number. No materialistic interests. The number is powerful if used wisely.

COMPOUND LIFE NUMBER 31
Sometimes connected with high intellect or genius. Similar to 30 – a number of seclusion and the recluse. Fixed attitudes. Very sensitive to the natural world.

Right: compound life number 27 reflects originality and imagination and will bring karmic rewards through hard work.

USING DIAGRAMS FOR INTERPRETATIONS

COMPOUND LIFE NUMBER 32
Magical connections. Communication – working exceptionally well under pressure, but likely to be influenced by others' opinions. Remaining true to their beliefs will avoid loss.

COMPOUND LIFE NUMBER 33
Similar to 24 with more chance of success and fortunate partnerships. Guard against over-confidence when things are going well. Doubly lucky number.

COMPOUND LIFE NUMBER 34
Linked to 25

COMPOUND LIFE NUMBER 35
Linked to 26

COMPOUND LIFE NUMBER 36
Linked to 27

COMPOUND LIFE NUMBER 37
Sexy, magnetic, charming individuals. Fortunate in all romantic relationships, friendships and partnerships. Sometimes volatile in close relationships.

COMPOUND LIFE NUMBER 38
Linked to 29

COMPOUND LIFE NUMBER 39
Linked to 30

COMPOUND LIFE NUMBER 40
Linked to 31

COMPOUND LIFE NUMBER 41
Linked to 32

COMPOUND LIFE NUMBER 42
Linked to 24

COMPOUND LIFE NUMBER 43
Failure and disappointment. Linked with revolution and anarchy.

Left: number 33 is a doubly lucky number, but over-confidence should be avoided when things are going well.

USING DIAGRAMS FOR INTERPRETATIONS

Above: number 51 – the number of the warrior, suggesting achievement and success against all odds.

COMPOUND LIFE NUMBER 44
Linked to 26

COMPOUND LIFE NUMBER 45
Linked to 27

COMPOUND LIFE NUMBER 46
Linked to 37

COMPOUND LIFE NUMBER 47
Linked to 29

COMPOUND LIFE NUMBER 48
Linked to 30

COMPOUND LIFE NUMBER 49
Linked to 31

COMPOUND LIFE NUMBER 50
Linked to 32

COMPOUND LIFE NUMBER 51
The fighter. The number of the warrior, suggesting achievement and success against all odds. Warning of enemies and danger. A number of wisdom.

COMPOUND LIFE NUMBER 52
Linked to 45

USING DIAGRAMS FOR INTERPRETATIONS

HEART NUMBERS 2
Sensitive, kind and vulnerable. There is an overwhelming desire to give to and care for others. Gifted healers, these people need security in their relationships and will work hard to achieve this.

HEART NUMBERS 3
Eternal optimists. Underneath the wit and humour is hidden a need for love. These are the people who are afraid of being rejected or left out and this is hidden from the rest of the world. They are encouraging parents, very reliable partners and wonderful friends.

HEART NUMBER 4
Hating the limelight of the centre stage, these are sensitive shy people who need a happy secure family. Born home-makers – usually they live in comfortable spotless surroundings – they love to build and improve their own homes.

HEART NUMBERS 5
Relationships can bring problems. Although these people have many friends, they also dislike authority and restriction in any shape or form. Their interest in philosophy may lead them to others with similar ideas and take to studying on their own quite late in life.

HEART NUMBER 6
Love is the key for these people whose concern is always for others. They put the needs of others first and always want to ensure the happiness of their partners and children. They do need praise and support to feel totally confident and relaxed about themselves.

HEART NUMBER 7
Happiest with their own company, they are mystics and philosophers and close relationships make these people uncomfortable.

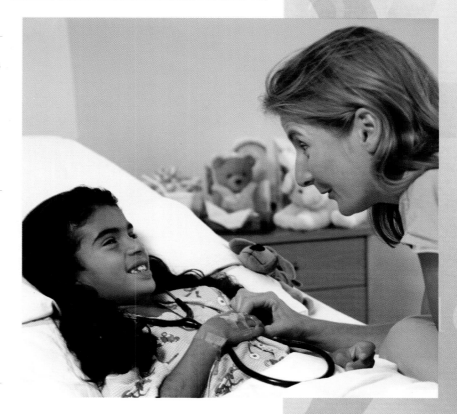

Above: people with heart number 2 have an overwhelming desire to give to and care for others.

USING DIAGRAMS FOR INTERPRETATIONS

HEART NUMBER 8
Being in control makes these people feel secure. They can be domineering and organising – quite unable to relax and enjoy themselves. Can be moody when things aren't going their way.

HEART NUMBER 9
Give these people a problem to solve and they will be happy. Good organisers who love variety and challenge. Men may be restless and they are often good mechanics. These individuals have a tendency to insist on knowing what makes everything work – including relationships – which may cause problems.

HEART NUMBER 11
Once these people have decided where they are going and what they want, they cannot be persuaded to change. They need to feel wanted in spite of having a strong inner faith in themselves and their abilities. Ideals and principles are most important to them.

HEART NUMBER 22
Provided they give up any unrealistic dreams and work hard, these people can achieve anything they want. Charming companions – friends and relations love their company – they love to create the feeling of peace and harmony which, for them, is perfection.

Right: heart number 9 people are great problem solvers and love a challenge; men who have this number are often good mechanics.

USING DIAGRAMS FOR INTERPRETATIONS

Destiny Numbers

What is meant by 'destiny' numbers? Destiny numbers give information about a person's lifestyle and they relate to the commonly used name. When you calculate the destiny number of a name, use the full name. Together with the life number, this is one of the most important numerological calculations.

Destiny Numbers
1. *Personal identity*
2. *Conscious mind*
3. *Personal creativity*
4. *Logical thought*
5. *Learning*
6. *Theory*
7. *Limits of material world*
8. *Unconscious mind*
9. *Completion*

How to discover your Destiny Number

1. Find the numerical values of the letters in your full name and add them up, reducing to a single number.

2. Find the interpretations of the compound and single numbers.

3. Circle the single number on the energy grid and find the keyword for that number.

This is your **destiny number** and the circled square on the grid shows your destiny energies.

You can now synthesise the interpretations to make a complete analysis of your destiny number.

PAUL DICKERSON

PERSONALITY NUMBER = 6
HEART NUMBER = 7

7 + 6 = 13
1 + 3 = 4

Destiny Number 1

The people who relate to number 1 are always in charge – they are often considered bossy and sometimes tyrants. It is the number of leadership and these are people who hate being told what to do, being restricted and having to do what anyone else wants. They need to be the centre of attention – always. They are creative and love being the head of a large family. Relationships, although sexy and passionate, can be demanding and often affairs outside marriage or partnership are essential for these individuals.

USING DIAGRAMS FOR INTERPRETATIONS

DESTINY NUMBER 2
Well balanced, diplomatic and level-headed, these people create an atmosphere of calm in situations which no-one else seems able to control. Tactful negotiators, they are protective, keep their own counsel and do well in any job which requires these special talents. They are very loving and solicitous towards their family, but sometimes people find them patronising and over-considerate, and feel stifled by their attention. Relationships with the right partner can be passionate and successful.

DESTINY NUMBER 3
These are people who are determined to succeed in everything they decide to do. They are bright, cheerful and enthusiastic. When they are disappointed they can hide their feelings well behind quick wit and humour – so they make excellent actors. Although they love an audience, in personal relationships they are often lonely. In spite of needing to feel free, they desperately need to feel loved and cherished so do best with someone who will be happy to share their life. In loving relationships the sense of humour of this number is at its best. They have an unexpectedly flirtatious side which long-term relationships do not always satisfy.

DESTINY NUMBER 4
These are the people who spend their lives making lists. They are organised, methodical, efficient and systematic, making wonderful employees. Often they remain single throughout their lives, preferring the company of animals to close human relationships. They love to be involved in the community. If they marry, it is often for money. In relationships they will be faithful and predictable, but often need to feel in control.

Below: destiny number 3 people are determined to succeed in everything they decide to do. They are bright, cheerful and enthusiastic.

USING DIAGRAMS FOR INTERPRETATIONS

DESTINY NUMBER 5

Hard workers and excellent communicators, these people need change and variety to function well. They can become restless if required to concentrate on something they don't want to do and will soon escape from restrictions. Travel is vital to them and they move around a great deal. In relationships they love the thrill of the chase and find it hard to settle down with one partner. Although they have a strong sex drive, once the novelty wears off, they may move on quickly to the next romantic challenge.

DESTINY NUMBER 6

If they are self-employed, these people are often very successful in their own business. Hard taskmasters and perfectionists, they channel much energy into making money. They love to make others happy and content. Although these people are naturally attractive, their attitude to relationships is not always traditional. Often they don't notice when their partner needs love and affection. Although they are loving and faithful, they demand that their partner always looks good, which can put extreme pressure on the relationship.

Left: destiny number 6 people are often very successful in their own business, being hard taskmasters and perfectionists.

USING DIAGRAMS FOR INTERPRETATIONS

DESTINY NUMBER 7

Dreams of saving the planet, adopting a cause or embracing philosophy and mysticism are the key traits of this number. These people work well with others in a teaching or helping role. Although they are loving partners, they do need to feel free to do what they want. Music is important and they are renowned for their mood swings. In relationships, these attractive, physical, sexy people never seem to be able to relax completely and a partner is often left wondering.

DESTINY NUMBER 8

The business tycoon who works hard, takes on responsibility and is very successful often exhibits complete intolerance of the failings of others. Because they are so well organised in their career, their home lives often leave much to be desired. In relationships they can appear to be cold and are too ambitious to put the effort into long-term partnerships or marriage. Often the partner will look for love and affection elsewhere. In a relationship which also includes work or business they can be happy and contented – even passionate.

Below: music is an important part of the lives of people with destiny number 7.

Destiny Number 9

Good at making split-second decisions, they can appear rash and impatient to others, although these are well organised people. They want everything now and are unwilling to wait for anyone or anything. They like variety and challenge. In partnerships they will be exceptionally loyal and can be a tower of strength in any crisis. It is often hard for these people to form one-to-one relationships because they prefer to be part of a group. They are often warm and caring, understanding that personal love is important and will be extremely generous partners.

Destiny Number 11

Communication is important here and often these are the people who will be found associated with journalism, TV or films. Excellent writers and teachers, they are naturally creative and good at making influential friends. They make very good leaders and have high ideals which they rarely compromise. They have much to learn about one-to-one relationships because they love the power of control. If they do give time to concentrate on a partnership they will be successful, but it is hard for them not to put their career first.

Destiny Number 22

Personal magnetism makes these individuals stand out in a crowd and attract attention. They often use this to get ahead in their career or job. Idealistic as well as practical, these people are usually successful provided that they don't take their exceptional abilities and talents for granted. In relationships they are usually very successful people and have many admirers. Essentially they are faithful to one partner although this may not always be reciprocated.

Left: Communication is important to people with destiny number 11 and often these are the people who will be found associated with journalism, TV or films.

USING DIAGRAMS FOR INTERPRETATIONS

Karmic Numbers
1. *Personal resources*
2. *Conscious mind*
3. *Action*
4. *The concrete*
5. *The senses*
6. *Abstract thinking*
7. *Bridge to the spiritual world*
8. *Timeless space*
9. *Karma – reward*

KARMIC NUMBERS

What is a karmic number?

If the theories of karma are to be believed, each person has lessons to learn within their lifetime. It is a matter of choice whether or not an individual wishes to improve their own karma. Sometimes what is revealed through the karmic number is not what a person wants to be confronted with and this should always be considered when looking at karmic numbers.

To find the karmic number add together the life number derived from the date of birth to the destiny number derived from the name. These numbers should always be considered together with any other interpretations about a person in an analysis of character.

PAUL DICKERSON

LIFE NUMBER = 4
DESTINY NUMBER = 6
4 + 6 = 10
1 + 0 = 1

3	6	9
2	5	8
①	4	7

HOW TO DISCOVER YOUR KARMIC NUMBER

1. Add together your life, personality, heart and destiny single numbers. Reduce them to a single digit and you will discover your karmic number.

2. Circle that single number on the energy grid and find the keyword for that number.

This is your **karmic number** and the circled square on the grid shows your karmic energies.

If you join the numbers you have circled with energy lines and read the energy interpretations, you discover the basic energies available to you.

Year Number 2

Try to keep the resolutions you made last year. Things may not seem to be moving fast enough for your liking, so try to develop patience and establish a basic harmony. It is not a good idea to take major decisions this year. You should attempt to take other people into account and work with them which means developing a certain tact and diplomacy. Other people may have their own ideas which they don't wish to change.

Good health trends, opportunities for travel and possibly difficulties in romantic affairs are likely this year. If you are considering a change of address, it would be wiser to wait until next year. It is possible that you decide to take up a new activity which is creative this year – if so, it may assume importance in years to come. Work hard to establish partnerships of all kinds this year. Hold fast to your long-term aims and try hard to relax whenever the chance presents itself.

Year Number 3

This may be the year when you reap the rewards of past efforts. There is a high level of activity with the likelihood of being in the public eye. If you are single, social gatherings, happiness, fun and romance are possible. Don't let negative people spoil things for you. During a 3 year all partnerships and friendships are good. Maybe you will decide it is time for a new image. Keep a careful eye on your diet.

A burst of creativity could bring new opportunities. Take them – you should be full of enthusiasm and feeling optimistic. You can have a very productive and happy year if you are determined to make the best of everything. The number 3 is generally considered to be lucky and many people will have wins through gambling this year.

Year Number 4

Material needs are important this year so get down to business and work hard. This is not the year to put too much energy or time into your social life. Lay the foundations for the future by making plans, clear away anything unnecessary and prepare to start afresh.

This year a great deal of your time will be spent thinking about money – you may well have extra unexpected expense. Be careful to save whenever you have the chance because financial stability will be important. Don't get too greedy or obsessed by money concerns.

Plans which include a logical order and the setting of goals to achieve success will be important. Target-setting and making long-needed improvements at home will be satisfying, and many people will find themselves becoming the centre of attention for a change, improving both status and appearance. The summer months will show progress in all your affairs; watch your health.

Year Number 5

This year is halfway through the 9-year cycle and may herald a decision to change many things. Moving home or job, travel and speculation will be fortunate, especially during the summer months. Think through any decisions carefully because now is the best time to change anything about your life which you don't feel comfortable with. Stick to your aims and goals – do not let others stand in your way

This is the number of communication. You will have a full diary – lots of meetings, telephone calls, e-mails and a renewed popularity with the opposite sex will make you feel that you are caught up in a whirlwind. Go everywhere, meet everyone and experience everything this year. People will be taking notice of you and you may start some new venture or get promotion at work.

Expect the unexpected throughout the year and remember to take every new opportunity and chance for advancement.

Year Number 6

The main emphasis this year is on your personal affairs. Sort out all domestic and emotional issues which you have not yet attended to. It could be matrimonial problems or difficulties with friends. Perhaps there are outstanding legal matters to be resolved. Friendships which have been under a cloud could improve during this year.

It is not always possible to clear up all difficulties and this year could be one when you experience personal loss or separation.

This year people who are single could consider marriage; you will feel a need to be content and settled in relationships and the greatest issues are related to balance and harmony.

This is a very important year for love.

Year Number 7

Step back this year and have a period of relaxation and rest. This will give you a chance to take stock of your life and consider yourself more than you normally do. Concentrate your energy on your own welfare; take care of your own health before that of your family for once.

A short period of time away on your own or periods of meditation may be what you need this year.

Because travel is highlighted this year you may find yourself taking short trips away and, if your job is involved with travel, you will have an especially good year at work. Unexpected opportunities to travel may take you by surprise and you should find these will allow you the chances you need to learn new things.

Do not spend too much time worrying about the material aspects of your life this year – it is more a time for philosophical or spiritual issues. Don't be surprised to discover a new interest in mysticism, magic and spirituality. If you are working in this field already, this will be an extremely rewarding year. Even if such matters are of no interest to you, you may find yourself spending more of your time and energy helping others throughout the year.

Left: year number 6 is a good year for love and an ideal time to consider marriage.

DAYS, MONTHS AND YEARS

YEAR NUMBER 8

Rewards for past efforts will become apparent during this year. They may be slow to arrive, but they will certainly arrive eventually. Many people will have financial and business successes; advancement of all kinds and promotion at work are highlighted throughout the year.

Romantic issues may be highlighted – this is an important year for the emotions and is sometimes referred to as a karmic year. If in the past you have been unwise or reckless, this is the time when things will catch up with you. Some may experience losses and a new feeling of instability.

Finances will be emphasised which may mean that some people get a welcome windfall – others may be made redundant or have financial problems. During this year property matters will be important – many people may be involved in buying or selling property.

Elderly relatives may become more part of your life and older people in general can find that they have important roles to fulfil.

Right: year number 8 could signify a property purchase or indeed a sale.

DAYS, MONTHS AND YEARS

APRIL   KEYWORD: DETERMINATION

1st	Dignity
2nd	Idealist
3rd	Order
4th	Initiative
5th	Consequence
6th	Experiment
7th	Belief
8th	Conscience
9th	Excess
10th	Daring
11th	Star
12th	Awareness
13th	Iconoclast
14th	Tradition
15th	Definition
16th	Comedy
17th	Purpose
18th	Defence
19th	Control
20th	Challenge
21st	Commitment
22nd	Establishment
23rd	Security
24th	Protection
25th	Substance
26th	Cultivation
27th	Self-sufficiency
28th	Steadfastness
29th	Image
30th	Overload

MAY	KEYWORD: CAUTION		
1st	Insight	17th	Bottom line
2nd	Observation	18th	Activity
3rd	Reality	19th	Persuasion
4th	Support	20th	Expression
5th	Awakening	21st	Vision
6th	Fantasy	22nd	Energy
7th	Devotion	23rd	Attention
8th	Truth	24th	Opinion
9th	Courage	25th	Boldness
10th	Isolation	26th	Protection
11th	Imagination	27th	Dedication
12th	Maverick	28th	Innovator
13th	Appeal	29th	Quicksilver
14th	Self-confidence	30th	Independence
15th	Dreamer	31st	Cutting edge
16th	Flair		

DAYS, MONTHS AND YEARS

JUNE — KEYWORD: DELAYS

1st	Being seen	16th	Investment
2nd	Problem-solver	17th	Force
3rd	Expression	18th	Security
4th	Expertise	19th	Spark
5th	Brilliance	20th	Appeal
6th	Visionary	21st	Rapture
7th	Entertainer	22nd	Exaltation
8th	Influence	23rd	Enchantment
9th	Insistence	24th	Wizard
10th	Laughter	25th	Receptivity
11th	Boundaries	26th	Stamina
12th	Optimism	27th	Development
13th	Adventure	28th	Stimulation
14th	Confrontation	29th	Dreamers
15th	Seduction	30th	Motivation

88 DAYS, MONTHS AND YEARS

JULY   KEYWORD: TRAVEL

1st	Emancipation
2nd	Unconscious
3rd	Commemoration
4th	Group
5th	Showman
6th	Desire
7th	Revelation
8th	Pragmatism
9th	Wonder
10th	Duality
11th	Privacy
12th	Persuasion
13th	Opportunity
14th	Storyteller
15th	Realisation
16th	Rising tide
17th	Career
18th	Conviction
19th	Controlled movement
20th	Ups and downs
21st	Controversy
22nd	Fluctuation
23rd	Resolution
24th	Instability
25th	Exploits
26th	Herald
27th	Decision
28th	Winner
29th	Assessment
30th	Tangible presence
31st	Human portrait

DAYS, MONTHS AND YEARS

AUGUST   KEYWORD: CHANGE

1st	Style	17th	Power
2nd	Versatility	18th	Endurance
3rd	Quest	19th	Surprise
4th	Guiding light	20th	Secrets
5th	Composure	21st	Privacy
6th	Unique happenings	22nd	Experience
7th	Double agent	23rd	Precision
8th	Role player	24th	Examination
9th	Tower of strength	25th	Exhibitionist
10th	Voice	26th	Support
11th	Validation	27th	Ideals
12th	Convention	28th	Language
13th	Long odds	29th	Structured action
14th	Mortal mirror	30th	Rock
15th	Royal command	31st	Public appearance
16th	High voltage		

DAYS, MONTHS AND YEARS

SEPTEMBER   KEYWORD: ACTION

1st	No nonsense		16th	Spirited energy
2nd	Egalitarian		17th	Perseverance
3rd	Breaking the mould		18th	Mystery
4th	Builder		19th	Appearance
5th	Idealist		20th	Manager
6th	Fate		21st	Taste
7th	Success		22nd	Restless drive
8th	Purist		23rd	Breakthrough
9th	Demand		24th	Wanderer
10th	Private goal		25th	Satirist
11th	Dramatic choice		26th	Patient practice
12th	Fearless crusader		27th	Ambiguous hero
13th	Passionate care		28th	Heartbreaker
14th	Perceptive critic		29th	Reaction
15th	Mastery		30th	Glaring truth

DAYS, MONTHS AND YEARS

OCTOBER   KEYWORD: FULFILMENT

1st	Top dog
2nd	Candour
3rd	Trendsetter
4th	Survival
5th	Just cause
6th	Good life
7th	Defiance
8th	Romance
9th	Intuition
10th	Economy
11th	Ease
12th	Grand gesture
13th	Dangerous enemy
14th	Moderation
15th	Personal magnetism
16th	Essential judgement
17th	Precarious balance
18th	Personal leadership
19th	Independence
20th	Influence
21st	Difference
22nd	Allure
23rd	Conflict
24th	Detail
25th	Form
26th	Organisation
27th	Impulse
28th	Research
29th	New ideas
30th	Overseer
31st	Attentiveness

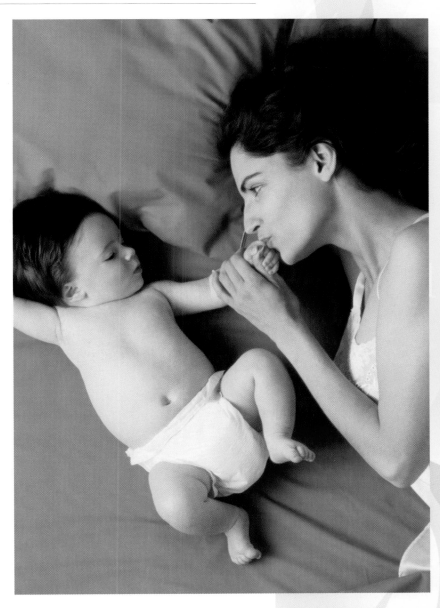

DAYS, MONTHS AND YEARS

NOVEMBER KEYWORDS: ACTIVITY BEHIND THE SCENES

1st	Onslaught		17th	Bridge
2nd	Transformation		18th	Temperament
3rd	Endurance		19th	Reformer
4th	Provocateur		20th	Struggle
5th	Actuality		21st	Elegance
6th	Vigour		22nd	Liberator
7th	Discovery		23rd	Irreverence
8th	Borderline		24th	Contention
9th	Temptation		25th	Sustained effort
10th	Metamorphosis		26th	Distinction
11th	Underground		27th	Excitement
12th	Charisma		28th	Lone wolf
13th	Commentator		29th	Instigator
14th	Investigator		30th	Measured attack
15th	Encounter			
16th	Authority